SpringerBriefs in Applied Sciences and Technology

W0193237

For further volumes:
http://www.springer.com/series/8884

Yasuhiro Suzuki · Rieko Suzuki

Tactile Score

A Knowledge Media for Tactile Sense

 Springer

Yasuhiro Suzuki
School of Informatics and Sciences
Nagoya University
Nagoya
Japan

Rieko Suzuki
Face Therapie Co., Ltd.
Tokyo
Japan

and

Graduate School of System Design
and Management
Keio University
Yokohama
Japan

Additional material to this book can be downloaded from http://extras.springer.com.

ISSN 2191-530X   ISSN 2191-5318   (electronic)
ISBN 978-4-431-54546-0   ISBN 978-4-431-54547-7   (eBook)
DOI 10.1007/978-4-431-54547-7
Springer Tokyo Heidelberg New York Dordrecht London

Library of Congress Control Number: 2013947783

Printed on acid-free paper

Springer is part of Springer Science+Business Media (www.springer.com)

Preface

Engineering application of a *Tactile Score* has been studied in human interfaces, human engineering, robot engineering, and other areas. In these fields, the central problem has been to recreate a real tactile sense mechanically. For example, there have been many studies related to fabric textures and measurements in material engineering. There are many studies in terms of measurements of the tactile sense and the development of tactile textures. However, conventional research has not been conducted to find a method for a temporal and spatial combination of the tactile sense.

A prototypical application of a spatiotemporal combination of the tactile sense is massage. Massage has been used for many years in medical treatments, in beauty therapies, and in other ways, but little scientific research on massage has been done. One of the challenges in research is that a method for describing massage has not been developed. For that reason, we are proposing a method, or measure, for describing massage: tactile score. The tactile score denotes the pressure intensity, the size of the contact area, and the rhythm of strokes. These three elements can be considered in the context of music scores by regarding pressure intensity as the pitch of a tone, the size of the contact area as the number of tones (i.e., whether a single note or a chord), and the rhythm of strokes as musical notes.

In this book, after a short review of general research on massage and our own research in the field, we introduce the basics of the tactile score and tactile composition, then we show the results of the investigation of a kind of massage called FaceTherapie™, which has been applied to more than 4,00,000 subjects. By using an approach from the field of psychology, the semantic differential method, we derive the basic techniques of massage, and by combining these techniques we are able to create various kinds of massage. Finally, we introduce projects for the application of the tactile score. With emotional engineering, we develop facial equipment for beauty treatment by using the tactile score, and also a transformation method for extrapolating from a music score to a tactile score. We hope this book will help to open a new area of tactile research and will bring about a new appreciation of the joy of the tactile sense.

Yasuhiro Suzuki
Rieko Suzuki

Acknowledgments

The authors express their gratitude for the collaboration of the following individuals and organizations: Dr. Junji Watanabe (NTT Communication Science Laboratories, Nippon Telegraph and Telephone Corporation) in the language of the tactile sense; Ms. Satoko Inaba and Ms. Mariko Umemura for experiments in massage; Prof. Takashi Maeno and Prof. Yasutoshi Makino (both at the Graduate School of System Design Management, Keio University) and Prof. Shigeru Sakurazawa (Future University Hakodate) in the development and discussion of haptic engineering; and to Prof. Fuminori Akiba (Nagoya University) for enlightening discussions. We thank Mrs. Rie Taniguchi (Nagoya University) and Springer Japan for their Editorial Support. Parts of our Research were supported by JSPS KAKENHI Grants-in-Aid for Scientific Research (B) No. 23300317 and (C) No. 24520106, which we gratefully acknowledge.

Contents

Chapter 1
Introduction

Abstract In this section, I introduce studies and designs related to Tactile score. The scientific research of massage has been conducted centered on medicine and psychology, and those researches show the effectiveness of human hand power. In this chapter, we introduce the tactile sense technologies, design methods of tactile sense and show summary of researches related to massage.

Keywords Tactile sense technologies · Design methods of tactile sense · Review of tactile sense researches

1.1 Tactile Sense Technologies

Almost all parts of our bodies are covered with skin. If something touches the skin, we can feel it. The skin connects our outside and inside that means our mind through tactile sense but some parts such as eyes are not covered with skin. For example, an eye is a part to see and an ear is a part to hear. All the same, these parts connect our outside and inside. Speaking in broad terms, all the sense organs such as skin, eyes, and ears are the membrane which connects our inside and outside.

When something touches the membrane, we feel the touch of it. For example, it is a baby's soft cheek, mellow music, gentle word, delicious food, and tender solicitude. It has various meanings such as touch, sound, image, taste, and concern. We can feel them as the sense being touched on the membrane, in other words, tactile sense.

Tactile sense (skin sensibility) consists mostly of the entire body among membranes that wrap our bodies. Tactile sense (skin sensibility) is unique organ which can't "close" among the senses consist our membranes. We can cover our ears, close our eyes, hold our nose, and close our mouth but not close tactile sense (skin sensibility). The importance of this tactile sense has become recognized and the research of tactile sense has been conducted in various areas.

Y. Suzuki and R. Suzuki, *Tactile Score*,
SpringerBriefs in Applied Sciences and Technology,
DOI: 10.1007/978-4-431-54547-7_1, © The Author(s) 2014

Tactile sense has been an interesting subject for basic science such as psychology, psychophysics, cognitive science and so on; and recently it also has been of interest to engineering and design. In engineering, technologies of tactile senses have been developed in the virtual reality, robotics, ergonomics and so on; where one of main subjects is "how to regenerate tactile sense mechanically" (for example [1]) and necessary and desirable applications of such technologies have been explored. I would like to leave the review in regard to enormous researches on tactile engineering to another books. In the past, the application of the tactile engineering to entertainment such as an integration of a tactile sense device to a video game controller and the application to communication technology such as applications for mobile phones have been mainly conducted.

For example, A. Chang et al. proposed the *ComTouch*, which is a device that augments remote voice communication with touch by converting hand pressure into vibrational intensity between the users in real-time. They used 24 people to observe possible uses of the tactile channel when used in conjunction with audio. By recording and examining both audio and tactile data, they found strong relationships between the two communication channels. Their studies show that the users developed an encoding system similar to that of Morse code, as well as three original uses: emphasis, mimicry, and turn-taking [2].

1.2 Design Methods of Tactile Sense

In the product design or manufacturing, tactile sense is an important factor; for example in the product design of electronic equipment such as a *smart phone* or *iPad*, tactile sense is a key factor for designing (for example [3]).

In textile science or Ergonomics, what is often called as Fabric hand or handle has been developed; it is defined as the human tactile sensory response towards fabric, which involves not only physical but also physiological, perceptional and social factors [Pan, 4]. "*Peirce in 1930 [4] first proposed to evaluate fabric hand based on physical measurement data. Since then, there have been several attempts to use instruments to measure fabric hand. All these efforts climaxed in 1970 when S. Kawabata and his co-workers in Japan developed a KES-FB system [5, 6] for fabric hand evaluation [4, P49]*". Hence we are able to design fabrics with reference to the evaluation of Fabric hand.

Studies of material dimension in human engineering express material textures and subjective distances of several materials as quantitative subjective data and extract potential factors. For example, J. Watanabe has verbalized images of touch from many tactile materials by using onomatopoeia and made an image map of tactile sense [7] and Chang have suggested a tactile circle based on the idea of a color circle regarding the application of haptic technology to communications [2].

1.3 Researches of Massage

There are enormous researches of underlying physiology and psychophysics about tactile sense. They are related to touch receptors and perceptions of touch in low order. Therefore, its relation to perceptual activities in high order such as pleasure-displeasure has still remained as a challenging research. As I mentioned in the Sect. 1.1, how to reproduce realistic tactile sense is the central project in touch engineering, but the applications of that technology are not many.

Various studies of material textures have been conducted regarding classifications of each material to the material texture dimension, however there are very few studies of the material dimension when those materials are combined temporally or spatially. In the case of temporal change in the touch material, for example, is massage. The tactile sense made with hands in the massage changes over time.

For example, a massage has been a state of art technique of tactile sense for a long time; since a massage affects our mental and physiology, it has been used in broad area and its effects have been investigated in (rehabilitation) medicine, psychiatry, the art of cosmetic treatments and so on, and it has been shown that a massage improves functional recovery in rehabilitation, brings realization and improves the condition of skin. And also tactile sense has been used in education, training of self-awareness and so on.

1.4 Massage in Medical Treatments

Billhult et al. [8] reported that when breast cancer patients received light pressure effleurage massage, the deterioration of NK cell activity was decreased during radiation therapy, and heart rate and systolic blood pressure were lowered.

And in [9] it is suggested that massage may be more effective than simple touch in decreasing pain and improving mood immediately for patients with advanced cancer because they may be touch-deprived by reason of social isolation or fear of causing harm. These findings support offering massage for immediate symptom relief and considering the potential therapeutic benefits of simple touch, which could be provided by family members or hospice volunteers, as an adjunct to usual care.

Also [10] showed a significant improvement in the eczema in the two groups of children following therapy, but there was no significant difference in improvement shown between the aromatherapy massage and massage only group. Thus there is evidence that tactile contact between mother and child benefits the symptoms of atopic eczema.

In [11], when patients with burn received 5 weeks of massage therapy, the measures including the pain, itching, and state anxiety were collected on the first and last days of the study period. The authors observed that massage therapy reduced all these measures from the first to the last day of this study. In most cultures, massage treatments are used to alleviate a wide range of symptoms. Although health professionals agree on the use of non-pharmacologic method for patients with burns, these applications are not yet common.

These studies are all alike in the point that patients' pain and fear is relieved by the action to be touched.

However [12] pointed out that massage alone or the application of compression after a single session of lymphatic massage was ineffective for reducing lymphedema in women with arm lymphedema secondary to breast cancer. This study shows a negative effect of massage to lymphedema.

1.5 Facial Massage

Timur Tashan and Kafkasli [13] reported that a 15-minute massage applied with almond oil during pregnancy reduced the development of striae gravidarum, but using bitter almond oil had no effect on this in itself. It is recommended that pregnant women be informed about the positive effects of massage applied with almond oil early during their pregnancy. This study shows a good effect of massage to skin. Tomoko et al. [14] concluded that the facial massage might refresh the subjects by reducing their psychological distress and activating the sympathetic nervous system. Ejidu [15] reported that both facial and foot treatments were equally effective in subjects' vital signs and reducing subjective levels of alertness during the interventions, with face massage marginally better at producing subjective sleepiness. Khanna and Datta Gupta [16] gave a warning that facial massage may have some adverse effects; although there are several subjective benefits with facial beauty treatment, there may be immediate side-effects, such as erythema and edema, as well as delayed problems, such as dermatitis and acneiform eruption, in about one-third of patients.

As seen from the above, the results of the massage research show some effects to various subjects, however one common factor in all of these studies is the power of human hands. Abraham Verghese [17] says the medical technology that will make the most progress in the next decade is the power of the human hand.

References

1. Eric M, Richard S, Stephen PS (eds) The Oxford handbook of philosophy of cognitive science. Oxford University Press, Oxford, 2010
2. Chang A, O'Sullivan C, (2008) An audio-haptic aesthetic framework influenced by visual theory. Springer-Verlag, Berlin Heidelberg, pp 70–80
3. Takeo Co. Ltd, (ed) Haptic: takeo paper show 2004. Asahi Shinbun, Tokyo, 2004
4. Peirce FT (1930) The 'handle' of cloth as a measurable quantity. J Text Inst 21:T377–T416

5. Kawabata S(1980) Examination of effect of basic mechanical properties of fabrics on fabric hand. In: Mechanics of flexible fiber assemblies, NATO Advanced Study Institute Series. Sijthoff and Noordhoff, Germantown, pp 405–417

6. Kawabata S (1980) The standardization and analysis of handle evaluation, 2nd edn. The Textile Machinery Society of Japan, Osaka

7. Hayakawa Tomohiko, Matsui Shigeru, Watanabe Junji (2010) Classification method of tactile textures using onomatopoeias. J Virtual Reality Jpn 15(3):487–490

8. Billhult A, Lindholm C, Gunnarsson R, Stener-Victorin E (2009) The effect of massage on immune function and stress in women with breast cancer–A randomized controlled trial. Auton Neurosci Basic Clin 150(1–2):111–115

9. Kutner JS, Smith MC, Corbin L, Hemphill L, Benton K, Karen Mellis B, Beaty B, Felton S, Yamashita TE, Bryant LL, Fairclough DL (2008) Massage therapy versus simple touch to improve pain and mood in patients with advanced cancer. Ann Intern Med 149:369-379

10. Anderson C, Lis-Balchin M, Kirk-Smith M (2000) Evaluation of massage with essential oils on childhood atopic eczema. Phytotherapy Research Phytother Res 14:452–456

11. Parlak Gürol A, Polat S, Akçay MN (2010) Itching, pain, and anxiety levels Are reduced with massage therapy in burned adolescents. J Burn Care Res 31(3):429–432

12. Maher J, Refshauge K, Ward L, Paterson R, Kilbreath S (2012) Change in extracellular fluid and arm volumes as a consequence of a single session of lymphatic massage followed by rest with or without compression. Support Care Cancer 20:3079–3086

13. Tashan Sermin Timur, Kafkasli Ayse (2012) The effect of bitter almond oil and massaging on striae gravidarum in primiparaous women. J Clin Nurs 21:1570–1576. doi:10.1111/j.1365-2702.2012.04087.x

14. Tomoko H, Shingo K, Chihiro T, Mayumi N, Koichiro O (2008) The facial massage reduced anxiety and negative mood status, and increased sympathetic nervous activity. Biomed Res 29(6):317–320

15. Ejidu Anna (2007) The effects of foot and facial massage on sleep induction, blood pressure, pulse and respiratory rate: crossover pilot study. Complement Ther Clin Pract 13:266–275

16. Khanna Neena, Gupta Siddhartha Datta (2002) Rejuvenating facial massage–a bane or boon? Int J Dermatol 41:407–410

17. Abraham Verghese (2011) A doctor's touch. http://www.ted.com/talks/abraham_verghese_a_doctor_s_touch.html

18. Pan N (2007) Quantification and evaluation of human tactile sense towards fabrics. Int J Des Nat 1:48–60

Chapter 2
Tactile Workshop

Abstract Tactile sense has been considered important for education, deep understanding of our selves, and sharing personal emotion caused by tactile sense with each other. We introduce tactile workshops and *"Haptica* bodyworkshop", which we have been developed.

Keywords Tactile workshop · Buruno Munarri · Somaesthetics · Richard Shusterman · Haptica · Bodyworkshop

There is a workshop with tactile sense paying attention to the change of the tactile sense and it also changes the human sense. For example, Italian designer, *Bruno Munari* educates children through the sense of touch in the tactile workshops [1]. In his workshop, various types of haptic materials are given to children and they express their emotion provoked by touching and combining these tactile materials.

Richard Shusterman has been proposed the somaesthetics, it is a new interdisciplinary field whose roots are in philosophical theory, somaesthetics offers an integrative conceptual framework and a menu of methodologies not only for better understanding our somatic experience, but also for improving the quality of our bodily perception, performance, and presentation. Such heightened somatic awareness and mastery offers benefits to many fields including design. Our experience of ourselves and in our world is always embodied, and it involves somatic responses and feelings that are typically unnoticed though they are unavoidable and indispensable for our proficient function. We need a proper feel for our tools in order to use them most effectively; and this includes the use of one's own body with using other tools. For the body is our indispensable tool of tools, the necessary medium of our being, perception, action and self-presentation in the world. By exploring the fundamental features of our embodied ways of engaging the world and transforming it through action and construction, somaesthetics can provide useful insights and experiential skills to help designers produce products and situations that provide more rewarding and pleasurable experience [2].

Y. Suzuki and R. Suzuki, *Tactile Score*,
SpringerBriefs in Applied Sciences and Technology,
DOI: 10.1007/978-4-431-54547-7_2, © The Author(s) 2014

He has been organizing bodyworkshop as a certified practitioner of Feldenkrais Method and a somatic therapist. He gives workshops on somaesthetics that include practical exercises and demonstrations, but also has experience in treating different cases of somatic disabilities.

2.1 Haptica Project

Authors (Suzuki and Suzuki) have been organizing the project focused on tactile sense of massage, haptica project, since 2002 [3]; our challenge has been how tactile sense is shared with others. In the most of massage, the tactile sense produced by massage is shared only between the one who gives massage and the one who receives it and it is difficult to share the tactile sense other than them. Hence, we have been doing bodyworkshops of tactile sense produced by massage and exploring the way to share the tactile sense with everyone.

2.1.1 Design of Bodyworkshop

In order to realize importance of tactile sense and reconsider it, we have developed bodyworkshops of tactile sense through giving facial massage; since most people are not interested in tactile sense in their daily lives, in every bodyworkshop we start with a pre-workshop, which urges people to tactile sense. Then we ask participants to make a pair with somebody and one person massages partner's face then exchanges the role and the person who received massage gives massage to the partner next.

2.1.2 Example of Pre-workshop

Through experience of bodysorkshops, we have learnt the importance of pre-workshop; in some bodyworkshops we omitted pre-workshops because the time was limited or the required style of workshop was different from our ordinal style, in such a case, every person tended to hesitate touching the partner's face and it took long time to start massage with a whole hand.

We have designed three types of pre-workshops;

(i) Play a game of tactile sense,
(ii) Do work with visual deprivation,
(iii) Create artworks related to tactile sense;

Fig. 2.1 Pre-workshop:
Tools for the game of tactile
sense, **a** heavy chopsticks,
30 cm long and 500 g, **b** light
chopsticks, 50 cm long and
about 3 g, **c** object to be
carried on a plate, these tools
were produced by Rieko
Suzuki and they were used in
the Haptica Bodyworkshop,
at the TFT Salon in
Koishikawa Tokyo

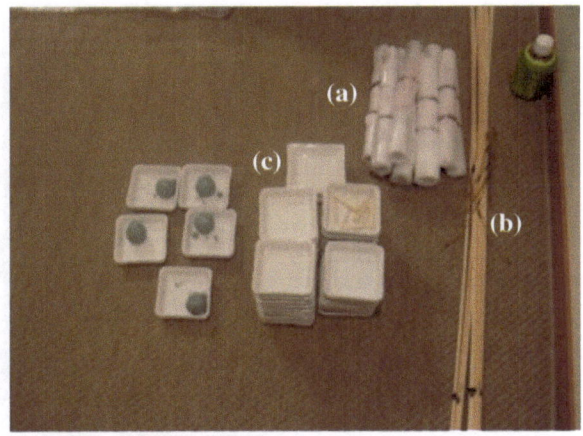

(i) Play a game of tactile sense: we have designed various games, for example we
    designed a "relay race" by carrying object one place to another; where a set of
    objects on plates and various types of chopsticks are prepared; the size and the
    weight of every object is different and they are put on plates, these plates are
    placed at intervals of about one meter on a line and different types of chop-
    sticks are placed beside each plate; a player requires to carry the first object on
    the first plate by the prepared chopsticks to the second plate and then carry the
    second object on the second plate to the third plate; when the player reach the
    last plate, loops back to the start point doing the same thing. Prior to Pre-work,
    two or more persons make a team, and they compete as a team in this relay
    game (Fig. 2.1).
(ii) In this pre-workshop: every participant puts a bandage over their eyes and eats
    a snack in a bowl with chopsticks; where snacks have various haptic feels of
    materials and they do not have strong favors in order to give concentration of
    participants' tactile sense, we ask participants to use chopsticks with visual
    deprivation so they have to concentrate on selecting and carrying a snack to
    their mouth without seeing or sniffing anything about the selected snack, then
    they have to concentrate their tactile senses of tongue or teeth on it to know
    what the selected snack is.

Participants can see and touch it, and also lie on it. These artworks were
produced by Rieko Suzuki (Haptica bodyworkshop in Honen-in Temple, Kyoto,
March 2005).

(iii) Before the pre-workshop: we created artworks for tactile sense and exhibited
    them in the workshop place. Participants can enjoy art exhibition not only
    seeing but also touching them; then every participant expresses impressions
    of each artwork by writing a poet, drawing a picture or creating a sculpture
    (Fig. 2.2).

Fig. 2.2 Pre-workshop with artworks: (*upper*) Participants are sitting around the artwork; (*lower*) an artwork for the pre-workshop, which is made of cotton and gel, and whose diameter is 1 M

2.1.3 Main Workshop

After the pre-workshop, we proceed on to the main workshop that we massage face; in order to induce participants' concentrations to massage, the most effective way is to ask participants to make an "image" of massage; for example, looking up at the surface of the water from the bottom of the deep sea. After the workshop, we asked them to express the image of the way they massaged by drawing pictures and making sculptures with paper clay. We also asked those who were massaged to express the image form the massage in the same way. Then the pair can share the tactile sense just between them in the massage by exhibiting their images to the other participants (Fig. 2.3).

Fig. 2.3 Main workshop: (*upper*) from Haptica bodyworkshop in Honen-in Temple, Kyoto, (*lower*) from the invited workshop in the International Symposium on Multi-sensory Design, Nagoya University, 2006

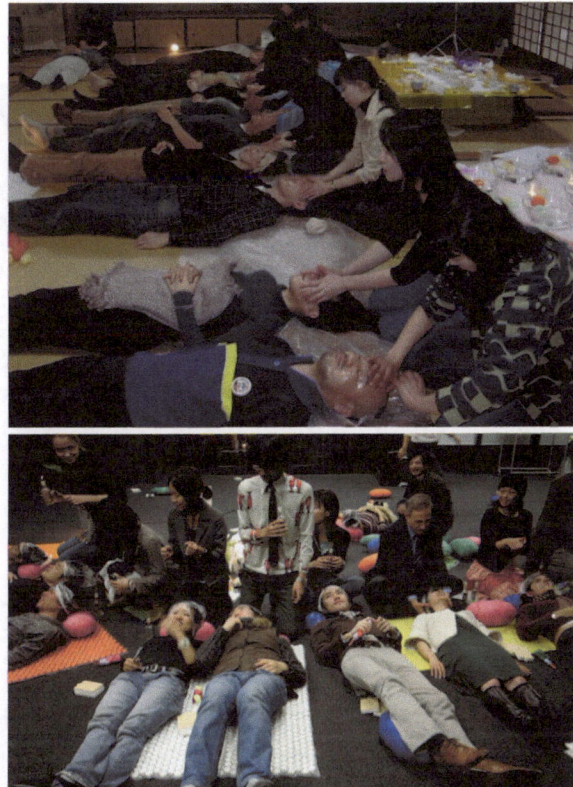

2.2 Bodyworkshops for Education of Children

We have developed bodyworkshop and explored the way to share the tactile sense of massage with every workshop participant and we found that expressing the tactile sense by creating artworks such as drawing a picture or making a sculpture with paper clay is effective to share the tactile sense. The most effective way was to require participants to create an image of the massage before doing the main workshop and in order to induce such image from participants, pre-workshop was useful. And it was important that after the main workshop, every pair those who were massaged to express the image and impression of the massage by showing their artworks (Fig. 2.4).

From experiences of bodyworkshops, we have realized the tactile sense reflects personality and mental status of mind, hence we believe that enrichment of tactile sense is an indispensable factor in education; as we mentioned Bruno Munarri had also pointed it out. Hence we have done workshops for children; in a elementary school (Hongo Elementary School, Tokyo), we explained the importance of tactile sense by giving a short lecture as the pre-workshop and did the main workshop,

Fig. 2.4 (*upper*) A participant of the workshop who was massaged expressed its tactile impression by drawing the painting and share with other participants by showing it, where a person next to her was her partner and she gave massage to her. (*lower*) the sculpture made of paper clay and a poet that were produced by a participant; they expressed the image of massage by the participant

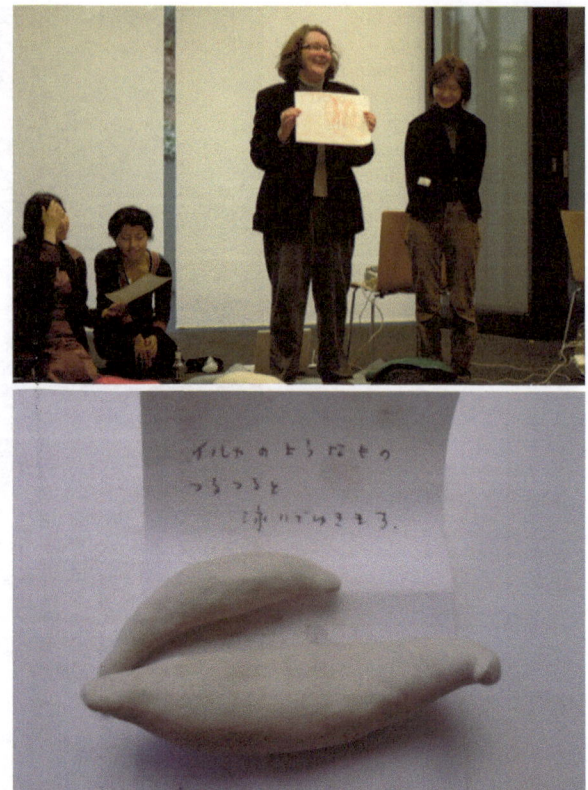

where we met few students who were not able to touch their partner; they did not smile or laugh very much from the begging of the workshop, however as the workshop was excited, they attempted to touch their partner and finally they could massage partner's face with their whole palms then they made a big laugh. After the workshop, every student expressed his or her impression of massage by drawing a picture, where a student expressed his feeling by drawing and writing the kanji character; he explained that he felt sadness from the massage and it looked like seeing a blue moon in the sky all alone at night (Fig. 2.5).

We have mainly developed bodyworkshops with massage but we also have organized workshop by using artworks for tactile sense, where we have created artworks to induce the tactile sense by toughing them, for example we created artworks with using various sizes of balls; they were placed in a dark tunnel and participants entered into the tunnel and crawled along it; the size of balls increased from small to large, near the entrance of the tunnel, the size of balls were 2 or 3 cm and the size of balls increased, and at the exit of the tunnel, there was a large ball whose diameter was about 1 meter; hence participants could sense the different sizes of balls with their whole body (Fig. 2.6).

Fig. 2.5 Haptica bodyworkshop in Hongo Elementary School, Tokyo; (*upper*) a student massages his partner; we asked every student to close their eyes while giving and receiving the massage in order to concentrate on tactile sense. (*lower*) After the workshop, students express the impression of tactile sense by drawing a picture

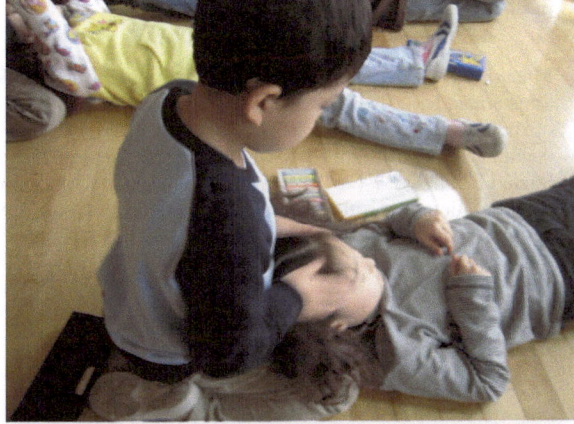

Fig. 2.6 Bodyworkshop at Aichi Children's Center, this artwork (sensing small and large with your whole body) was produced by Rieko Suzuki

References

1. Munari B (1985) I laboratory tattili. Edizioni Corraini, Bologna
2. Shusterman R (2013): Somaesthetics. In: Soegaard M, Dam M, Friis R (eds) The encyclopedia of human-computer interaction, 2nd edn. Aarhus: The Interaction Design Foundation. Available online at http://www.interaction-design.org/encyclopedia/somaesthetics.html
3. Suzuki R, Suzuki Y (2013) How to "share" the tactile sense?, a putative approach 302—308, BookFrontiers in Artificial Intelligence and Applications.In: Intelligent interactive multimedia systems and services, vol 254. pp 302–308

Chapter 3
Experiments Relating to Massage

Abstract Tactile sense has been investigated in neuroscience, molecular biology, cognitive science, psychological physics and so on. These previous researchers have not examined higher order cognition and its mechanisms such as massage. In this chapter, we show our results on experiments about cognition of massage.

Keywords Cognition of massage · Psychological physics · Semantic differential method

3.1 Psycho-Physical Experiment of the Tactile Perception

Existing studies on tactile perception have centered on the generation of tactile stimulation (i.e. receiving end) but not much of giving end, i.e. how to touch has not been discussed. Thus, we have conducted a psychological experiment [1] on the change in the sensibility corresponding to different touching manners.

Right index and middle fingers of 11 subjects were stimulated with abrasive papers by the examiner, and the differences of coarseness that the subjects were able to detect (discrimination threshold) were monitored. Two were taken from the pool of abrasive papers whose average grain sizes were 1, 3, 30, 5, 9, 12, 40 μm and used as standard and comparison stimuli (Fig. 3.1). The number of test specimens was 56, which was all combinations, including that of identical grain size, were made. During the course of the experiment, the subjects were blindfolded and the examiner touched their right index and middle fingers simultaneously with the abrasive papers in the test specimens. Then, the subjects selected coarser ones [2].

Generally, we use the cumulative normal distribution function as a fitting function (Sigmoid curve) and we obtain the discrimination threshold as the stimulus intensity whose dictation rate is 50 %. Since the fitness function of 1, 3, 30 and 40 μm were not sigmoid curves, we could not obtain discrimination thresholds but could obtain when 5, 9 and 12 μm.

Y. Suzuki and R. Suzuki, *Tactile Score*,
SpringerBriefs in Applied Sciences and Technology,
DOI: 10.1007/978-4-431-54547-7_3, © The Author(s) 2014

Fig. 3.1 Psychophysical
experiment, on the test
specimens there are abrasive
papers as standard and
comparison stimuli: A special
glove restricts flexibility of
fingers other than the index
and the middle fingers. The
examiner moves the test
specimens

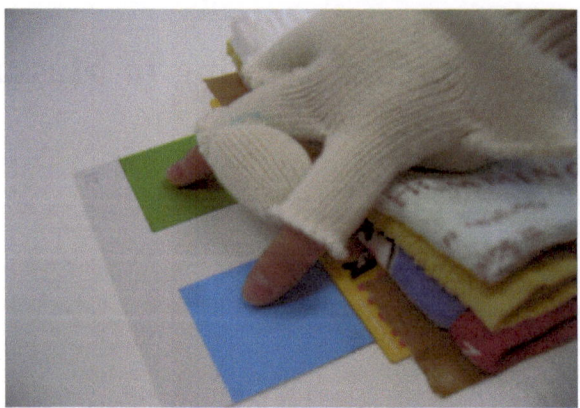

- Condition 1: The examiner mechanically moves the test specimens (to an electronic metronome).
- Condition 2: The examiner actively moves the test specimens.

In every grain size, discrimination thresholds of condition 2 were less than condition 1. Based on the analyses on the discrimination thresholds and Weber ratio (discrimination threshold/standard stimulus, Fig. 3.2), the discrimination sensitivity under Condition 2 was higher than Condition 1 (Table 3.2), suggesting that the smaller difference could be perceived when the subjects were touched actively, even when the same set of stimuli were used.

T test of difference in averages of condition 1 and 2 indicates statistical significance (T value was $2.973 > 2.920$: degree of freedom was 2, two-tailed test and significance level was 10 %). The average of miss-discrimination (in 56 test specimens) of condition 1 and 2 were 6.17 and 5.27, respectively. We confirmed that this difference had statistical significance, where T value was $1.928 > 1.812$: degree of freedom was 10, two tailed test and significance level was 10 % (Table 3.1).

Fig. 3.2 Weber ratios of
psychophysical experiment
with condition 1 and 2; the
horizontal axis denotes
standard stimulation and the
vertical axis denotes Weber
ratios

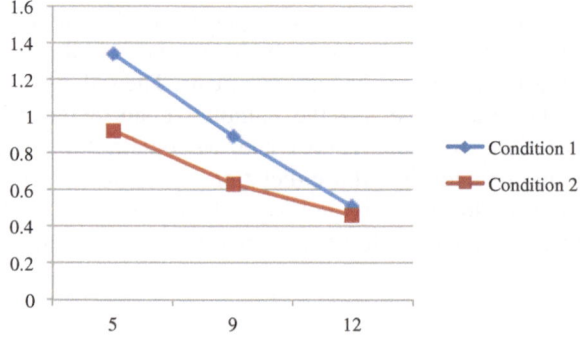

Table 3.1 Discrimination threshold

Standard stimulus	5	9	12	Average
Condition 1	6.697	8.0168	6.1193	6.9443
Condition 2	4.597	5.7352	5.5769	5.3030
Difference	2.100	2.2817	0.5423	1.6403

3.2 Psychonomics Approach

An examiner who is not a professional massage therapist touched the tactile stimuli and then performed circular massage on the cheeks of the subjects based on the image of the stimuli [4]. The tactile stimuli were balls with smooth, rough and fluffy tactile sensation (Fig. 3.3).

The examiner made sure that the hand movements were identical after touching any of the stimuli (Fig. 3.4). The subjects filled out questionnaires to give the impression of each massage. An examiner massaged the stimuli for one minute, and then massaged the subjects for two minutes. Subjects gave the impression of each massage by choosing an integer scaled betweens−3 and 3 for each pair of adjective.

Perception experiment of massage by the Semantic Differential (SD) method [3] and the psychological technique for impression analyses was conducted to investigate higher-order tactile cognition. In the SD method, the connotative meaning of concepts was measured by asking a respondent to choose where his or her position located in a scale between two bipolar adjectives. We used the following adjectives (Table 3.2).

We then examined the result using principal component analysis. The first principal component was the characteristics of strength of touch contribution was 33.27 % (strong–weak, positive–negative), the second principal component was impression of touch contribution was 14.39 % (hard–soft, gentle–fear, sleepy–wide awake, intense–quiet and desirable–not desirable, pleasant–unpleasant).

Fig. 3.3 Balls that give tactile stimuli: from left to right tactile sensation of smooth, fluffy and rough

Fig. 3.4 The examiner
performs circular massage on
the cheeks of the subject
based on the image of the
stimuli through touching a
ball

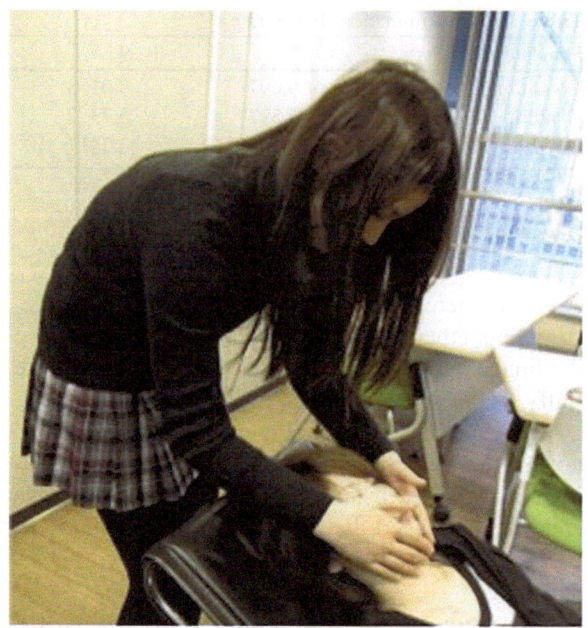

We obtained statistical significant pairs of adjectives in the questionnaires by
the analysis of variance and multiple comparison test then examine which tactile
stimuli corresponded to balls with smooth, rough or fluffy tactile sensation. The
result of the Multiple Comparison tests suggested that significant difference
between smooth and fluffy and also rough and fluffy. The difference in tactile
impressions gave subtle difference in massage, which was perceived by the subjects, despite the identical hand movement. Based on the above, it has been
suggested that active massage could heighten tactile reception with the perception
of subtle difference in the tactile stimuli.

Table 3.2 Pairs of adjectives
used in SD method analysis

Warm	Chilly
Hot	Cold
Heavy	Light
Hard	Soft
Pleasant	Unpleasant
Strong	Weak
Gentle	Fear
Sleepy	Wide awake
Quick	Slow
Intense	Quiet
Positive	Negative
Painful	Not painful
Desirable	Not desirable

3.3 Related Works and Discussion

It has been pointed out "the way of touching" with enhancing the perception of tactile sense; For example, the sheet metal inspectors in automobile industries know that the knitted work gloves help the perception of surface undulation [6]. Sano et al. have found three mechanisms of touch enhancing on work gloves. The first one is a lever mechanism of knitted gloves. The second one is a buckling phenomenon of the glove, which generates the tactile stimulus to the articular joint. The third one is the noise-mediated improvements, namely a stochastic resonance, which enhances the detectability of a weak stimulus [6]. And they have proposed a device for enhancing tactile perception of surface undulation. This device, which we call a "tactile contact lens," is composed of a sheet and numerous pins arranged on one side of the sheet. Our experimental results show that a small bump on a surface can be detected more accurately through this device than by bare fingers and than through a flat sheet.

They also analyzed the phenomenon provoked by this simple device and suggested two causes of this phenomenon. One is a lever-like behavior of the pins, which converts the local inclination of the object surface into the tangential displacement on the skin surface. The other is the spatial aliasing effect resulting from the discrete contact, by which the temporal change on the skin surface displacement is efficiently transduced into the temporal change in the skin tissue strain. The result of the analysis is discussed in relation to other sensitivity-enhancing materials, tactile sensing mechanisms, and tactile/haptic display devices [5].

In the related works, *"how to touch"* is changed mechanically by using texture of fabric or behaviors of "pins" on the sheet of the tactile contact lens. In our former experiment, we examined the difference of tactile perception when we changed how to touch between the *"passive touch"* and *"active touch"*; in the passive touch experiment, an examiner touched materials in the equal time interval, while in active touch experiment, touched materials freely to examine their surface roughness and it was shown that the active touch enhanced the tactile perception. This result indicates how to touch changes the tactile perception and we may be able to enhance the tactile perception by the way of touching.

And in the latter experiment, we showed that images in the mind of a person who gave a touch would change the way of touching and the difference of images may be perceptible by a subject (a person to be touched). This is a preliminary result and it should be carefully examined in more detail to conclude it scientifically; while in a beauty salon or massage school, the importance of having such images in massage has been confirmed empirically; for example, it is difficult for new students to massage the face of the subject with strong pressure, they try to press the face hard however their power only reaches the arms but does not reach hands. In such a case, we give an image; at first we ask them to touch the face with their normal pressure and press it a bit harder as if they go down the stairs; by having the image of going down the stairs most students are able to press the face hard.

Not only in the scene of a beauty salon or school but also experiences in bodyworkshops, we have confirmed that it is important to have images on massage but these images change in both "how to touch" and "how to be touched", so it has been difficult to describe massages clearly when we teach and investigate them. It is difficult to treat such personal images by others, so we do not treat it directly but examine it indirectly by describing massages in detail, then we have developed such a method for describing them or tactile sense. We introduce the method in the next chapter.

References

1. Stevens SS (1957) On the psychophysical law. Psychol Rev 64(3):153–181
2. Inaba S, Suzuki Y (2007) Research on the perception of active stimulation in human tactile sense. Graduation thesis of School of Information Science, Nagoya University
3. Snider JG, Osgood CE (1969) Semantic differential technique: a sourcebook. Aldine, Chicago
4. Umemura Mariko, Suzuki Yasuhiro (2011) Presentation of several haptic stimuli using massage and psychological study on the discrimination sensitivity on those stimuli. Nagoya University, Graduation thesis of School of Information Science
5. Kikuuwe R, Sano A, Mochiyama H, Takesue N, Tsunekawa K, Suzuki S, Fujimoto H (2004) Sensors, 2004. In: Proceedings of IEEE digital object identifier: 10.1109/ICSENS.2004.1426219. vol 2, pp 535–538
6. Sano A, Tanaka Y, Fujimoto H (2010) 1P1-F11 three mechanisms of touch enhancing on work gloves : lever mechanism, buckling, and stochastic resonance. In: The robotics and mechatronics conference 2010, 1P1-F11, The Japan Society of Mechanical Engineers

Chapter 4
Tactile Score, "Syoku-fu™"

Abstract In order to describe and design tactile sense generated by such as massage, we have developed *Tactile score* with reference to Music score. We introduce the "history" of Tactile score, *Shoku-fu™* (*"shoku"* stands for tactile sense and *"fu"*, a score in Japanese); why and how we have obtained the Tactile score. Various types of Tactile scores have been proposed and applied not only to describe and design massage but also to compose music, Emotional engineering, haptic design and so on. In this chapter, we introduce basics of Tactile sense and how to describe massage by using it.

Keywords History of Tactile score · Basics of Tactile score · Describe massage · Syoku-fu

4.1 The Way to "Tactile Score"

Since when we began beauty therapists, we have been creating various massages and techniques. There were some massages that didn't satisfy customers. We learned massages that were symmetrical, regular, planar, constant rate, and without thinking about dimensions or changes of pressures at a beauty school where we had professional trainings. Those were massages with a focus on the press of acupressure points and the flow of lymph. Thus, when we opened our own beauty salon, we massaged customers as we learned at the school. However, no matter how much we made efforts, we didn't show much beauty effects or couldn't attract more customers. We occasionally got high beauty effects, but we didn't understand why we got such effects.

Things went on so for a long time, and we felt the limits of our abilities as beauty therapists and started thinking to go out of the business. Though we didn't understand the reason, there was a case that our massage showed a high beauty effect and a customer was very satisfied with it. We empirically knew the massage method that gave high beauty effects, and then we tried to pursue the method.

Y. Suzuki and R. Suzuki, *Tactile Score*,
SpringerBriefs in Applied Sciences and Technology,
DOI: 10.1007/978-4-431-54547-7_4, © The Author(s) 2014

Fig. 4.1 The tactile sense of massage with figures like hieroglyphics

From about this time, we started describing the tactile sense of our massage with figures like hieroglyphics (Fig. 4.1). Hereafter we refer to this figure as the "Tactile word". We use words in order to code an object and decode the words and the combination of them for replicating massages.

First we extremely slowed the regular massage speed that we learned at the school. The regular massage was mechanical and inorganic but we did contrary massage such as artistic and organic. We imitated calligraphy, and so regarded our palms as brushes and massage oil as India ink, then tried to massage as if we drew characters. We massaged customers' faces very slowly as if we connected dots as to calligraphy. Then customers got angry and claimed that the massage was unpleasant.

We thought that the reason this unsuccessful massage didn't give pleasure to customers was that it was too slow to convey the rhythm. Then, we tried to create a massage which we could feel a rhythm. We put a bit pressure with fingers and changed the speed to music in order to convey a rhythm through the massage. We used hole palms in the massage referred to calligraphy, next we massaged a face as if it was a stage and danced our fingers on the face. However, we received negative reviews again. Customers claimed that there was no tactile sense even grotesque incoming sensation. We thought we massaged them quite rhythmically, but we found that people who received massage didn't feel any rhythm.

All these new trials were unpopular with customers and the number of customers had fallen. So we were forced to get back to the regular massage we learned at the school. We went back to the beginning of creating new massage and felt at a loss.

Then we listed the elements of our own hands movements with tactile words in order to objectively see what consisted our massage. We noticed that the use of tactile words made us think the combinations of massage techniques and the order of massage elements objectively. We came to be able to do a kind of computing such as addition, subtraction, and assembly with assembling basic elements of the massage such as the differences among right and left, pressure, contact areas of palms, and speed as if we assemble the parts of the jigsaw puzzle.

Through a trial and error process using tactile words, we began to see the most important elements are the dimension, pressure, and speed in *Face Therapie*™ which is our massage technique. Also we discovered that we could construct a spatial massage with the combination of these elements. Although a face is three dimensional, a massage is consisted of two-dimensional movements over the surface. Multi-dimensional changes, such as the contact area, pressure and speed among others were perceived from there.

No matter how we move our hands in three dimensions, we can't go into between skins, therefore, massage is basically planar. Even though there is asperity on a face, hands are moved with being attached firmly to the face, in other words, the hand movement itself is planar to the face. Therefore, we began to see that we could make the massage spatial with changing the pressure and speed of the massage as well as the hand movement. Then, we moved into the massage that we focused simply on the pressure change attaching little importance to the hand movement. However, customers claimed that they felt funny.

We learned that a stroke was one of the important elements of massage from this failure. We also found that the important point was not any one element but the massage constitutive priority. We noticed that the origin and development of these tactile words were similar to that of music scores. Then we created and proposed the tactile score. The tactile score has been variously improved in basically the same frame since then.

The creation of the tactile score made it possible to record the massage that gives good feeling in detail. The description capability of the tactile words is low, so the reproduced massage differs among people. However, anyone can reproduce the massage that has almost the same texture with the tactile score, and we can also create complicated massages.

4.2 Tactile Score, Syoku-fu™

We apply the scoring way to the tactile note to describe massage [1, 2]. The pressure intensity is expressed as a staff. Two Kanji characters in Japanese express "Shoku fu"; "*shoku*" stands for tactile sense and "*fu*", a score.

(Figure 4.2a) We set the line sandwiched in between two upside-down triangles as the basic pressure, and then move it up and down to create a pressure variation, for example, in describing the pressure when we touch something important.

(Figure 4.2b) The whole note represents speed, and it also includes a movement of a stroke (Fig. 4.4).

Fig. 4.2 The "staff notation" of tactile score

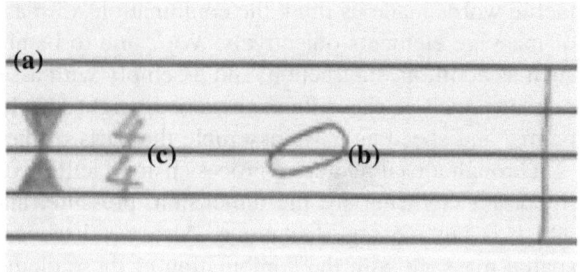

(Figure 4.2c) Based on our experiences, we found that we could give more comfort by beating time to the pressure and speed. Here the beat is quadruple time, but triple and double time are also acceptable (Fig. 4.4).

Next we number the areas of the palm to describe the size of the dimension (Fig. 4.3), in addition, we encode the spatial position and the movement of the stroke like a curve, line, dot, and each size of them like small, medium, and large as tactile steps like sol-fa of a musical score on a face (Fig. 4.3). Then, we classify the speeds of whole notes described above.

As well as whole notes in Fig. 4.2, you can also classify double notes, quarter notes, eight notes, and dots (Fig. 4.4). In staff notation of the tactile note, we define the third line as the basic pressure; the basic pressure is the pressure when we hold a baby or an expensive jewel very carefully. Hence, the basic pressure is not defined absolutely but may change from person to person or for different types of massage. For the tactile note, we define the pressure intensity as the difference in

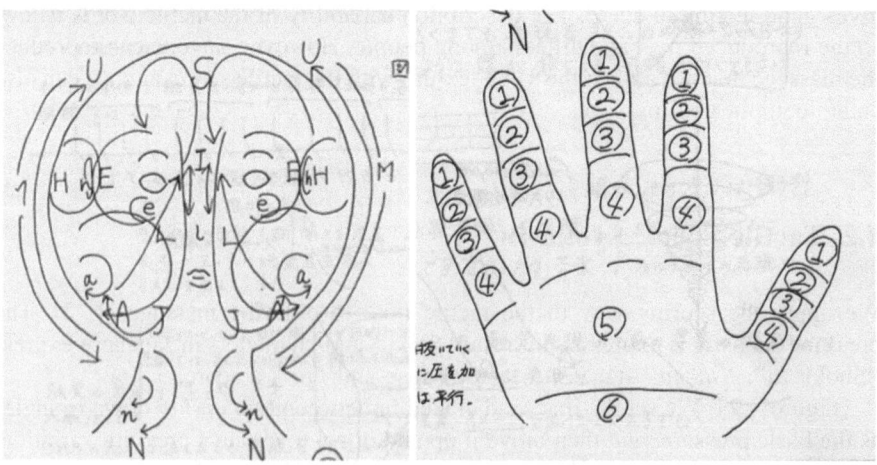

Fig. 4.3 *Left* Strokes of massage on a face; these strokes are obtained from massage experiences in aesthetic salons; strokes that pass uncomfortable areas have been excluded. *Right* Usage of parts of the hand

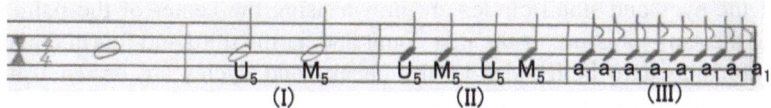

Fig. 4.4 Tactile score of whole, half, quarter, eighth notes

pressure from the basic pressure. We define stronger pressure as downward from the third line in the staff notation and weaker pressure as upward from the third line.

We also define the part of the hand and the kind of strokes used in massage (see Figs. 4.2, 4.3). For example, the fingertip to the first joint is 1, the second joint is 2, the third joint is 3, the upper part of the palm is 4, the center of the palm is 5 and the bottom of the palm is 6; when we use from a fingertip to the third joint, this is denoted as "1–3". For massage strokes, we analyze the method of massage, *Face Therapie*™ and extract strokes; we symbolize each stroke as A, a, N, n, etc. For example, the symbol A stands for the massage stroke of drawing a circle on the cheek. In this notation, for example, A_5 illustrates drawing a circle on the cheek with the center of the palm.

The tactile score in this contribution is the basic version in which each musical note denotes massage with both hands and we denote a gap in hand motion with a special mark above the staff notation;

1 denotes both hands moving at the same time,
2 indicate a small gap between hands and
3 indicate a large gap between hands.

Tactile score (as in the Fig. 4.5) has special symbols: 1 denotes both hands moving at the same time, 2 indicates a small gap between hands, and 3 indicates a large gap between hands, the Sulla like marks illustrate a unit component of massage, the integral-like marks illustrate releasing pressure, and the breath-like mark corresponds to a short pause in massage, much like a breath in playing music. Schematic expression illustrates the change of the pressure and contact area, where the size of each cycle illustrates the contact area and the solid line illustrates the pressure change.

And in this tactile score, at the first count in the beginning part, A_5, circles are drawn on both sides of the cheeks using the center of the palm with weaker pressure than the basic pressure, at the second count, the hands are moved to the

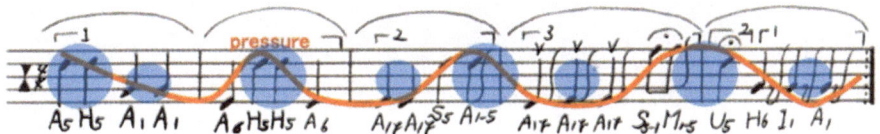

Fig. 4.5 Tactile score and special symbols

tails of the eyes and small circles are drawn using the center of the palm while keeping the same pressure as the first count and, at the third and fourth counts, the hands are moved to both sides of the cheeks and cycles are drawn using the fingertips with a stronger pressure than the basic pressure.

References

1. Akiba F, Suzuki Y (2007) Sociable aesthetics of bodily senses and the "haptica" projects. In: Proceedings of IIIrd Mediterranean Congress of Aesthetics, pp 13–16
2. Akiba F, Suzuki Y (2012) The computational aesthetics of the tactile sense and its significance for philosophical aesthetics. In: 22nd Biennial Congress of The International Association of Empirical Aesthetics (IAEA), National Taiwan Normal University, pp 558–561, 22–25

Chapter 5
Investigation of Massage by Using Tactile Score

Abstract In order to explore "hidden" language inside the massage technique, we investigate the construction of massage based on "basic" technique that we extracted from massages; and we examine each "image of massage" by using Semantic Differential method (SD method) and show massages composed of these basic techniques and having rules of composition.

Keywords Language of massage · Semantic differential method

By using a tactile score, we can analyze the standard massage used in *Face Therapie*™. We examined the method and confirmed that it can be broken down into 42 kinds of basic massage components [1, 2]. Through this investigation, we can describe various massages by combining these basic components. To characterize these basic components, we use the semantic differential (SD) method. We also asked the inventor of *Face Therapie*™ to be the respondent in our SD method analysis.

In this analysis, we used nine pairs of adjectives as follows (Table 5.1): the respondent chose an integer scaled between −3 and 3 for each basic component. We then examined the result using principal component analysis.

The first principal component was the characteristics of touch (soft–large, blow–wrap), the second principal component was the time variation of touch (disappearing–releasing), and the third principal component was the pressure change (heavy–sharp). By using these principal components, we classify 42 kinds of basic components into six groups, named I: light pressure, II middle pressure, III heavy pressure, IV light flow, V keen flow, and VI soft flow. (See Figs. 5.1, 5.2).

The characterization of basic components corresponds to the possession of drawing materials for basic motifs of massage. For example, when painting, we compose an artwork by using drawing materials to create a beautiful form. Tactile stimuli do not have visual or auditory forms for which we can judge their beauty. Hence we define the beauty of massage as comforts. For a beauty salon, it is required that massages can improve the skin condition or physical states of the body and attract customers; otherwise, the business fails. Hence, we define a massage that has kept a high client satisfaction level for ten years as a "beautiful

Y. Suzuki and R. Suzuki, *Tactile Score*,
SpringerBriefs in Applied Sciences and Technology,
DOI: 10.1007/978-4-431-54547-7_5, © The Author(s) 2014

Table 5.1 Pairs of adjectives used in SD method analysis

Soft	Hard
Light	Heavy
Large	Small
Sharp	Blunt
Disappearing	Remaining
Inhibitory	Releasing
Calm	Stable
Hollow	Blow
Dub	Wrap

Fig. 5.1 Six groups of 42 kinds of basic components, named I: light pressure, II middle pressure, III heavy pressure, IV light flow, V keen flow, and VI soft flow

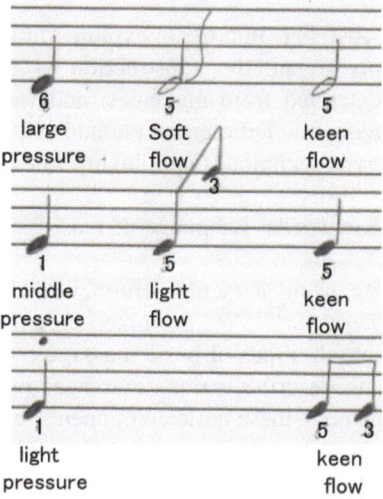

Fig. 5.2 Map of the 42 basic components in the principal component space, where the horizontal axis illustrates the first and second principal components and the vertical axis illustrates the third principal components

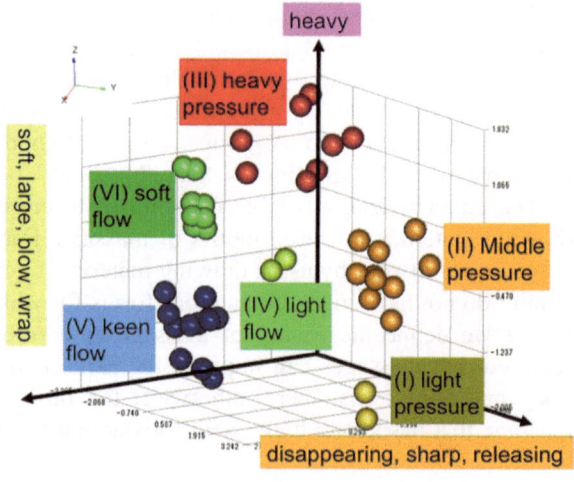

Fig. 5.3 The result of a time series of basic massages components, where each bidirectional arrows illustrates possible transitions between basic groups. Groups IV and V (indicated by cycles) are intermediate groups; they mediate transitions between groups

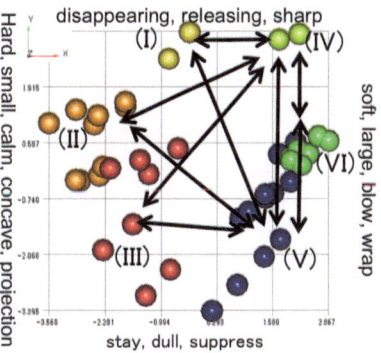

massage". We have studied various massages and found that comfortable massages are likely to be beautiful massages. The standard massage has been obtained through embodiment of such comfortable massages, so we analyze it. We described the standard massage using tactile scores and transformed it into basic components I to VI; and, we analyzed the massage as a time series of basic components. We found that the basic components of IV (middle pressure) and V (keen flow) are used as intermediate components; for example, for a massage starting from I (light pressure) to III (heavy pressure), since there is no direct transition path from I to III, it has to go through IV or V such as I → IV → III or I → V → III. This result indicates that in order to compose a massage, there is the principle of composing basic techniques (Fig. 5.3).

References

1. Suzuki Y, Watanabe J, Suzuki R (2012) Tactile score, a knowledge media of tactile sense for creativity. In: Watanabe T (ed) Smart innovation, systems and technologies vol 14, no.1, Springer Verlag, Berlin, pp 579–589
2. Suzuki R, Watanabe J, Suzuki Y (2013) Classification of technical primitive images in the massage described with tactile score, transaction of the virtual reality society of japan, vol.3 No. 13 (to appear)

Chapter 6
Method of Composing Massage

Abstract We have explored how to compose massages in beauty treatments by using Tactile score; it leads us to deep understandings about tactile sense and how to design massage. Tactile sense has *language* and we use it in our daily lives such as tapping someone, mother's gentle touch to a baby or children and so on. And we have obtained a method for composing Tactile score. In this chapter we introduce a method for "composing" massage by using Tactile score.

Keywords Method for composing Tactile score · Language of tactile sense

Though we created the tactile score, we couldn't receive positive feedback from customers regarding the complicated massage with the tactile score. They claimed that they felt a strange sensation and didn't quite understand that massage. We learned that a comfortable massage was more than just a complicated massage. Through this experience, we have been investigating how to compose Tactile score and we required deeper understanding on what is Tactile Sense [1].

The important thing for a comfortable massage is to develop a certain pattern. As the result of trial and error in creating a massage for high customer satisfaction, we discovered that the massage evaluated as comfortable was complicated and understandable. If a customer receives monotonous and repetitive massage, he/she gets bored, and when he/she is massaged with random pattern, he/she usually feels uncomfortable.

Experiences show that all the massages that give customers comfortable feelings and satisfaction have some kind of regularity. In general, if the composition is well organized, the complexity tends to be proportional to the comfort. The above experiences suggest the linguistic aspect of tactile score.

Y. Suzuki and R. Suzuki, *Tactile Score*,
SpringerBriefs in Applied Sciences and Technology,
DOI: 10.1007/978-4-431-54547-7_6, © The Author(s) 2014

6.1 Language of Tactile Sense

Tactile perception conveys different messages from speech language [1]. When one is patted on the shoulder once, he/she might think of accidental collision, yet when patted twice, it has meaning and he/she interprets it as someone has called. Also, mothers gently tap babies at steady rhythm in caressing; the steady rhythm evokes the sense of security in babies.

In other words, counts and rhythm are important in tactile perception; a single circular stroke could not be distinguished from a mere rubbing, while more than double strokes would be recognized as massage. So this "(more than) double strokes" is an "*alphabet* "of a *language of tactile sense* and a set composed of these alphabets is a "*word*"; a massage is composed of these words as if a *sentence*. A poet is composed of sentences and these sentences generate "rhythm"; likewise a poet, sentences composed of tactile sense words also generate rhythm.

As mothers' gently tap, steady rhythm added meaning and sense of security to massage so such steady rhythm would be considered as measures in music. Empirically, we have found that many subjects like massages composed of Tactile scores with quadruple measures, so when we compose a Tactile score, basic elements in 4 counts are used as one unit of massage.

By giving a rhythm on a tactile sense, we can create "*impressions*"; a rhythm of touching gives a "*theme*" on the impression provoked by tactile sense, where the theme is the expression through tactile sense such as small-large, fast-slow, line-curve and so on; if a sequence of massage strokes starts from small circles then moves to large circles and small circles again, a subject would feel small and large. Let us consider a "*jelly*" as a simplified model of someone's "face"; because it allows vibrations from outside caused by massage to pass through easily; if your vibrations to the jelly variously changes due to the combination of the strength of the touch, the width of the contact area and the speed of the hand then the movement of the jelly is changed and in some cases it generates rhythms; such rhythms would provoke various tactual stimulations. Persons who are touched/massaged are able to sense such various tactile stimulations as different "textures" likewise gentle, cold, solid, soft etc. Tactual textures of fabrics or materials have been investigated well as we have addressed in the Chap. 1; the most crucial difference of tactual texture in massage is that the texture emerges from "spatiotemporal stream" of tactile stimulations; in tactual texture of fabrics or materials, spatiotemporal combinations of tactile stimulations have not been considered very much.

We suppose that our image caused by tactile sense emerges from the temporal relationship; we always compare the tactile sense in the past and at the present. If we touch something hard, and then touch something harder, we regard the former as soft. So, the image will be determined by comparison of what/how we touch in the past and at the present; hence we can generate the tactile sense created by mother's hands by pairing such as hardness and softness and can generate a rhythm of tactile sense by designing the pair of tactile senses.

Fig. 6.1 A variety of pressure of hands and speed of hand movements. *1* Rub the subject hard, *2* Rub it little weaker than *1*, *3* Hold its muscle, *4* Hold it a little weaker than *3*, *5* Touch it lightly as if one touches downy hair

In the previous chapter, we have investigated massages and found that a massage has structural composition of tactile senses as if languages; if we tap someone two times to stop someone and if there is time lag between the first and second tap and the difference is 4 or 5 s, the person who was tapped would not be able to understand why you have tapped because both "tactically understandable" and "not understandable" tactile compositions exist in this case.

The important thing when we create a rhythm is an existence of a *theme*, that is, what a difference you want to tell the subject, and the same goes for pressure. If the pressure on the subject stays the same, she/he can't feel the pressure, but she/he feels the pressure if she/he is pushed hard during being touched softly. As for the speed, for example, when we move our hands back and forth in 10 s at first, next in 5 s, and then in 20 s, she/he feels a difference in speed (Fig. 6.1).

This makes a rhythm on the tactile sense with the theme of speed. In other words, a rhythm includes various differences such as hardness, softness, pressure, and speed. Combining the differences creates the expression of the tactile sense as if *weaving yarns* that are different colors of texture creates a rhythm of a woven fabric.

In addition, the rhythm of the tactile sense is created not only by tactual textures but also "*how to touch*". For example, when we express softness by a tactile sense, people try to touch softly; instead we touch with spreading and closing the fingers, and alternate these two moves. When we close the fingers, we place pressure upon the subject, and when we spread the fingers, we relieve the pressure so the "Softness" is produced by these moves. *Touching softly* and *touching soft tactual textures* are two different things; even if the subject is a pin holder, we feel as if it is smooth depending on the way of touching.

Fig. 6.2 An example of the
Tactile score of (i)

Fig. 6.3 An example of the
Tactile score of (ii)

Fig. 6.4 An example of the
Tactile score of (iii)

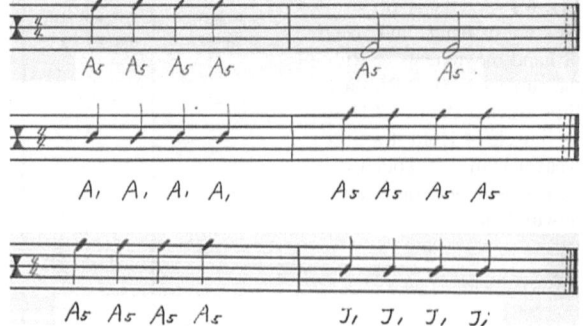

6.2 Exercise

Compose Tactile scores that express following themes

(i) weight,
(ii) size,
(iii) directions;

Hints: A "theme" in Tactile score is expressed by "differences"; for example, if the sequence of massage strokes starts from short lines then moves to long lines and short lines again, where the "differences" are expressed by the length of strokes so the theme of this Tactile score is "length."

Examples of Answers: The most important thing is how to create difference;

(i) In this Tactile score, "weight" is expressed by making difference in pressures; light pressure in the first measure and heavier pressure in the second measure (Fig. 6.2).
(ii) In this Tactile score, contact areas are enlarged in the second measure than first measure; in order to express "size", we make differences in the size of strokes by combining larger strokes than first measure (Fig. 6.3).
(iii) In this Tactile score, the type of strokes are changed from circular (A) to *J shaped* (J). In order to express the directions of strokes, we changed the types of strokes from circles to lines (Fig. 6.4).

the above Tactile scores are examples and you can freely compose Tactile score as you like; the way of checking whether the composed Tactile score is right or wrong is to massage yourself or someone following the composed Tactile score so you can confirm your Tactile score expresses its theme or not.

Reference

1. Suzuki R, Suzuki Y (2013) Introduction of facetherapie ("Facetherapie Nyumon," in Japanese), Syun-jyu-sya

Chapter 7
Future Tactile Sense

Abstract In this final chapter, we give another aspect of tactile sense and massage from the view point of Natural computing, NC; NC is a interdisciplinary research field relating to computer science, biology, chemistry and so on. An aim of NC is to understand nature as algorithm: in the previous chapter, we showed that massages can be regarded as basic techniques and its composition, where we are also able to regard basic techniques as "codes" and its composition as a "computer program," hence we can also consider a massage as a model of NC. We give a novel "platform" for investigating and designing massage.

Keywords Natural computing · Computational aesthetics

There is a room for further scientific research into massage however many kinds of massages have been improved and used all over the world from time immemorial because massage effects have repeatability to some extent. As we described in the chapter of Tactile score, the Tactile score was designed for increasing the repeatability of the massage effects and describing the massage method.

7.1 Creating Tactile Sense as Natural Computing

Natural computing is a research area in computer science and it aims to understand nature as algorithm; Natural computing has three main topics (i) computing with natural media (DNA, chemistry, slime mold, etc.), (ii) bio-inspired computing, (iii) Computational aesthetics. For example, DNA computing (computing with DNA) was proposed in 1960' and realized in 1994 [1] and has succeeded in Bio nano-technology such as technologies for constructing Molecular robotics [2].

As the definition of computing, the Church-Turing thesis has been accepted; in order to consider Natural computing, I expand this thesis to the basis of natural computing and proposed a definition of computing as "an order of codes and its execution" and we call an order of codes as a program; this program is processed not only by a computer but also by natural things such as a person, molecule, etc.

Y. Suzuki and R. Suzuki, *Tactile Score*,
SpringerBriefs in Applied Sciences and Technology,
DOI: 10.1007/978-4-431-54547-7_7, © The Author(s) 2014

In various researches of Natural computing, we have been extracted operable codes from natural system and create a program by ordering these codes as well as the computing with Turing Machine or computers.

For example T. Nakagaki et al. have been constructing the natural computing system with slime mold for searching the shortest path search in a maze [3]. Since slime molds dislike iron, they use iron to make obstructions and limit the movement of slime mold and locate the bait at the exit of maze, where the amount of bait is important; if the amount of bait is large, they does not need to search the shortest path and if it is small, they will have the total amount of bait before they change their shape, so in this natural computing system, the design of a maze made of iron and the amount of bait are codes.

We have extracted basic techniques of a massage and described them by using Tactile score; these basic massages correspond to "codes" and we can design an "order" of codes thus we can regard a massage as a kind of Natural computing, where an order of codes corresponds to a program and a person executes the program corresponds to a "computer" and the program is executed by a person and a massage is generated [4].

As described above, we introduce right to the point of the tactile score. We have been analyzing hand movements with making tactile score and creating new tactile senses in order to find a higher level of hand techniques.

Tactile score models after musical scores, so we are often asked the difference between tactile score and musical score. At first, we created the tactile score just to describe tactile sense made with hand techniques. Honestly, we didn't pay much attention to the relationship between the tactile score and music. However, we have a feeling that the more we research the tactile sense made with hands techniques, the more the relationship deepen.

The tactile score can describe not only massage, but also general tactile sense. We are making a tool that reproduces tactile score such as facial equipment with Professor T. Maeno and Y. Makino at System Design Methodology in Keio University. In this research development, I put together the tactile sense by reproducing the tactile score and music out of curiosity.

I think the music was Brahms' Symphony. When I listened to the music receiving tactile stimuli, I felt deep impression as if I were in a concert hall. It was a fresh surprise for me.

I was much inspired when the music and tactile stimuli were combined than with only listening to the music or receiving tactile stimuli from the facial equipment. We have taken a hint from this to develop the facial equipment that stimulates senses of hearing and touch at the same time. At first, we used copyright free sound sources, but now we can make music and tactile stimuli from tactile score at the same time because we recently began to see the method for directly putting a tactile score into music. We are on the verge of becoming possible to take musical scores as tactile scores though a trial and error process in researching the relationships between tactile sense and music.

Musical scores are one of mankind's great resources and the number and variety of them are really huge. We have been excited to imagine that the tactile world we have never experienced may lie inside musical scores. It's just like having been traveling to search the new tactile sense.

References

1. Adleman LM (1994) Molecular computation of solutions to combinatorial problems. Science 266(5187):1021–1024
2. Murata S, Konagaya A, Kobayashi S, Saito H, Hagiya M (2013) Molecular robotics: a new paradim for artifacts. New gener comput 31(1):27–45
3. Nakagaki Toshiyuki, Yamada Hiroyasu, To'th A'gota (2000) Maze-solving by an amoeboid organism. Nature 407:470
4. Suzuki Y (2013) Harness the nature for computation. In: Natural computing and beyond proceedings in information and communications technology Vol 6. Springer Verlag, Berlin, pp 49–70

Index

Y. Suzuki and R. Suzuki, *Tactile Score*,
SpringerBriefs in Applied Sciences and Technology,
DOI: 10.1007/978-4-431-54547-7, © The Author(s) 2014